Safe Lifting and Movement of Nursing Home Residents

by:

James W. Collins, PhD, MSME
Associate Director for Science
Division of Safety Research
National Institute for Occupational Safety and Health
Centers for Disease Control and Prevention
Morgantown, West Virginia

Audrey Nelson, PhD, RN, FAAN
Director
Patient Safety Center for Inquiry
James A. Haley Veteran's Administration Hospital
Tampa, Florida

Virginia Sublet, PhD, RPh
Senior Toxicologist
Oak Ridge Institute for Science and Education
Windermere, Florida

DEPARTMENT OF HEALTH AND HUMAN SERVICES
Centers for Disease Control and Prevention
National Institute for Occupational Safety and Health

Disclaimer

Ordering Information

To receive documents or other information about occupational safety and health topics, contact NIOSH at:

NIOSH-Publications Dissemination
4676 Columbia Parkway
Cincinnati, Ohio 45226-1998

Telephone: **1-800-35-NIOSH** (1-800-356-4674)
Fax: 513-533-8573
E-mail: pubstaft@cdc.gov

or visit the NIOSH Web site at **www.cdc.gov/niosh**

DHHS (NIOSH) Publication Number 2006-117

February 2006

SAFER · HEALTHIER · PEOPLE™

Table of Contents

Acknowledgments

The authors of this document would like to convey special thanks to the following individuals for their contribution and review of the document.

Contributors:

Laurie Wolf, MS, CPE
Certified Professional Ergonomist
BJC Health System
BJC Corporate Health Services
St. Louis, Missouri

Jennifer Bell, PhD
Epidemiologist
Centers for Disease Control and Prevention
National Institute for Occupational Safety and Health
Division of Safety Research
Morgantown, West Virginia

Brad Evanoff, MD, MPH
Associate Professor of Medicine
Director, Division of General Medical Sciences
Washington University-School of Medicine
St. Louis, Missouri

Arun Garg, PhD, CPE
Professor and Director of the Ergonomics Laboratory
Industrial and Manufacturing Engineering Department
University of Wisconsin – Milwaukee
Milwaukee, Wisconsin

Thomas W. Waters, PhD, CPE
Chief, Human Factors and Ergonomics Section
Centers for Disease Control and Prevention
National Institute for Occupational Safety and Health
Division of Applied Research and Technology
Cincinnati, Ohio

Reviewers:

Bill Borwegen, MPH
Director, Occupational Health and Safety
Service Employees International Union
Washington, DC

Guy Fragala, PhD, PE, CSP
Director of Compliance Programs
Environmental Health and Engineering, Inc.
Newton, Massachusetts

Pamela Hagan, MSN, RN
Chief Programs Officer
American Nurses Association
Silver Spring, Maryland

Anne Hudson, RN, BSN
Work Injured Nurses' Group (WING) USA
Coos Bay, Oregon

Joe Joliff, BS
Retired Administrator
Wyandot County Nursing Home
Upper Sandusky, Ohio

Joanna Sznajder, PhD, CPE
U.S. Department of Labor
Occupational Safety and Health Administration
Directorate of Standards and Guidance
Office of Physical Hazards
Washington, DC

The authors would also like to thank the nursing home owners, administrators, nurse managers, and safety and health professionals who participated in a series of focus groups and evaluated the content and format of this document.

Photo Credits:
Arjo Inc.
E Z Way Inc.
Liko Inc.
BHM Medical

Introduction

Introduction

This guide is intended for nursing home owners, administrators, nurse managers, safety and health professionals, and workers who are interested in establishing a safe resident lifting program. Research conducted by the National Institute for Occupational Safety and Health (NIOSH), the Veterans' Health Administration (VHA), and the University of Wisconsin-Milwaukee has shown that safe resident lifting programs that incorporate mechanical lifting equipment can protect workers from injury, reduce workers' compensation costs, and improve the quality of care delivered to residents. This guide also presents a business case to show that the investment in lifting equipment and training can be recovered through reduced workers' compensation expenses and costs associated with lost and restricted work days.

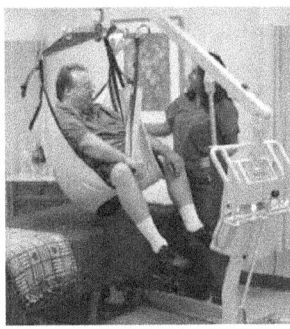 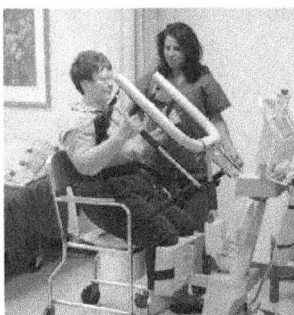 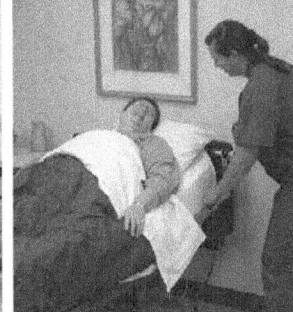

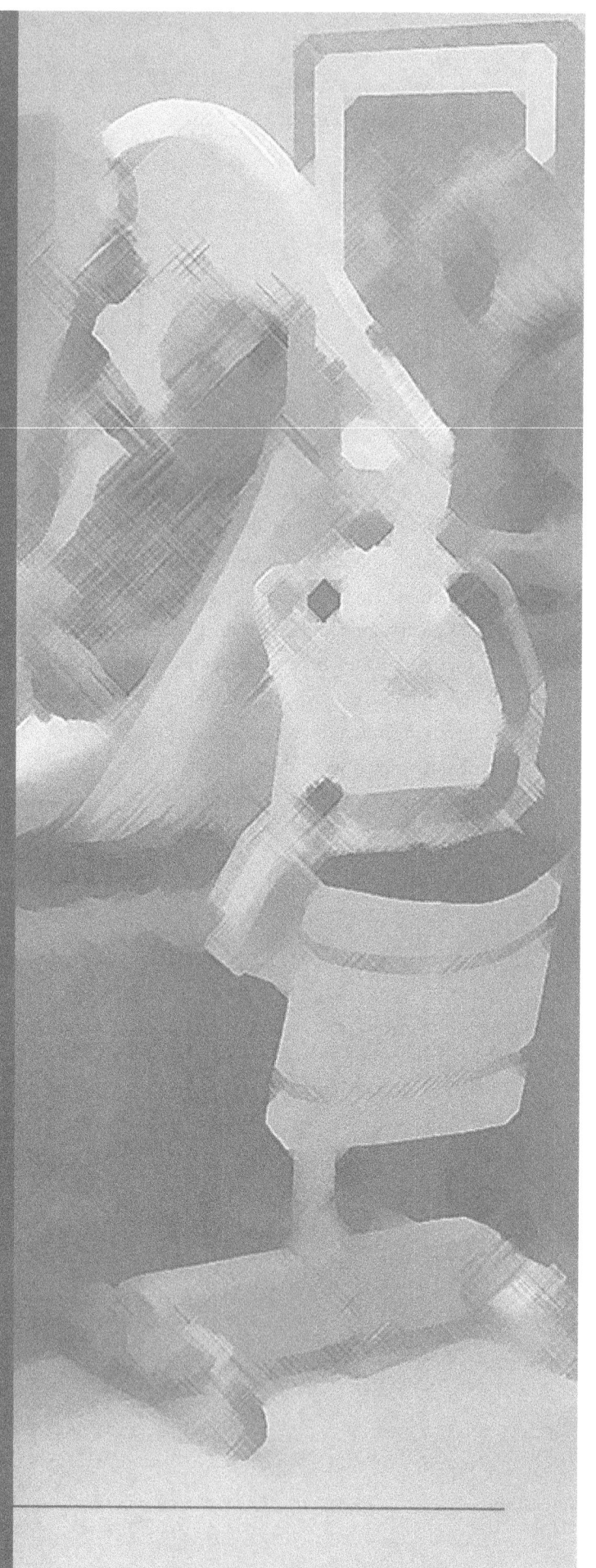

The Challenge of Lifting Residents in Nursing Homes

One of the major issues in nursing homes is the frequent heavy lifting and repositioning of residents that exceed the lifting capacity of most caregivers[1]. Numerous studies have shown that training caregivers how to use proper body mechanics to lift residents is not an effective prevention measure because lifting the weight of adult patients is intrinsically unsafe. Because of the trend towards shorter hospital stays, residents who are being transferred to nursing homes are becoming increasingly frail. Factors that contribute to the difficulty of lifting and moving a resident include the size and weight of the resident, combativeness, and propensity to fall or lose balance. In addition, performing resident transfers in the confines of small bathrooms and rooms cluttered with medical equipment and furniture works against the caregiver being able to use good body mechanics. When lifting or repositioning a resident in bed, the bed generally prevents the caregiver

from bending his/her knees to assume the proper posture for lifting. The forward bending required for many patient lifting and moving activities places the caregiver's spine in its most vulnerable position. Even under ideal lifting conditions, the weight of any adult far exceeds the lifting capacity of most caregivers, 90 percent of whom are female.

These conditions contributed to the 211,000 occupational injuries suffered by caregivers in 2003 (Bureau of Labor Statistics, 2003). Because of the rapidly expanding elderly population in the U.S., employment for nursing aides, orderlies, and attendants is projected to increase by 25% between 2002 and 2012, adding an estimated 343,000 jobs (Bureau of Labor Statistics, 2004). Due to the ongoing demand for skilled care services, musculoskeletal injuries to the back, shoulder, and upper extremities of caregivers are expected to increase.

---

[1] The term caregiver in this document is defined as nursing aides and orderlies, licensed practical nurses, registered nurses, therapists, and restorative aides employed in nursing homes.

Benefits, Cost, and Effectiveness of a Safe Resident Lifting Program

(1) What are the benefits of a safe resident lifting program?

The following benefits can be derived from a safe resident lifting program that includes mechanical lifting equipment, worker training on the use of the lifts, and a written resident lifting policy:

Benefits for Residents

- Improved quality of care
- Improved resident safety and comfort
- Improved resident satisfaction
- Reduced risk of falls, being dropped, friction burns, dislocated shoulders
- Reduced skin tears and bruises

Benefits for Employers

- Reduced number and severity of staff injuries
- Improved resident safety
- Reduced workers' compensation medical and indemnity costs

- Reduced lost workdays
- Reduced restricted workdays
- Reduced overtime and sick leave
- Improved recruitment and retention of caregivers
- Fewer resources required to replace injured staff

Benefits for Caregivers

- Reduced risk of injury
- Improved job satisfaction
- Increased morale
- Injured caregivers are less likely to be re-injured
- Pregnant caregivers can work longer
- Staff can work to an older age
- More energy at the end of the work shift
- Less pain and muscle fatigue on a daily basis

(2) How much will it cost to set up a safe resident lifting, handling, and movement program?

A 100-bed facility can expect to spend $25,000 to $30,000 on portable (not ceiling-mounted) mechanical lifts depending on how many residents in your facility require the use of a lift. As a general rule, one full-body lift should be provided for approximately every eight to ten non-weight bearing residents and one stand-up lift should be provided for approximately every eight to ten partially-weight bearing residents. The average cost of a mechanical lift can vary from $3,000 to $6,000 per lift. The average cost for a ceiling-mounted lift is approximately $4,000 per room. An effective combination of both floor and ceiling lifts is generally accomplished with a $50,000 to $60,000 investment per 100 bed facility.

(3) Can a nursing home recover the cost of implementing a safe resident lifting, handling and movement program?

Cost-benefit analyses demonstrate that the initial investment in lifting equipment and employee training can be recovered in two to three years through reductions in workers' compensation expenses (Collins et al., 2004; Tiesman et al., 2003; Nelson et al., 2003; Garg, 1999).

(4) How effective is mechanical lifting equipment in preventing injuries to caregivers?

Safe resident lifting programs can be highly effective in reducing a health care worker's exposure to heavy loads and awkward working postures that contribute to back and other musculoskeletal injuries. Research has shown that safe resident lifting programs reduce resident-handling workers' compensation injury rates by 61%, lost workday injury rates by 66%, restricted workdays by 38%, and the number of workers suffering from repeat injuries (Collins et al., 2004). Similar findings have been reported by other investigators (Tiesman et al., 2003; Nelson et al., 2003; Garg, 1999). Furthermore, this research has shown an increase in caregiver job satisfaction, and a decrease in "unsafe" patient handling practices performed. Nurses ranked lifting equipment as the most important element in a safe lifting program (Nelson et al., 2003). The increase in bariatric residents has also led lifting equipment manufacturers to develop equipment with higher lifting capacities to accomodate the special needs of some bariatric residents.

(5) What benefits will a safe resident lifting program have for nursing home residents?

Although some residents may be reluctant to try new lifting devices, studies have shown that the use of mechanical lifting equipment increases a resident's comfort and feelings of security when compared to manual methods (Zhuang et al., 2000; Garg and Owen 1992).

The findings from one study indicated that residents' acceptance of a safe lifting program was moderate when the program was first implemented but high at the end of the research study (Nelson et al., 2003).

Injuries to residents are also reduced because the mechanical lifts protect residents from being dropped. Anecdotal information indicates that a reduction in skin tears and bruises may result when residents are handled mechanically rather than manually (Garg, 1999).

Frequently Asked Questions about Safe Resident Lifting, Handling, and Movement Programs

(1) Is training in body mechanics alone an effective prevention measure for caregivers with regard to lifting and transferring residents?

Numerous studies have shown that training caregivers on how to use proper body mechanics to lift residents is not an effective prevention measure because lifting the weight of adult patients is intrinsically unsafe (Nelson et al., 2003).

(2) Are back belts an effective option for reducing the risk of back or other musculoskeletal injuries to caregivers?

The effectiveness of using back belts to lessen the risk of back injury among uninjured workers remains unproven and may give workers a false sense of security (Wassell et al., 2000; NIOSH, 1994).

(3) How can nursing home management motivate staff to use lifting equipment initially and maintain long-term commitment?

- Provide sufficient training on lift usage so that caregivers learn how to properly operate the equipment. Training should be provided to all newly hired caregivers and a plan should be in place to assess competency in use of the equipment, at least, annually.
- Post a graph to show caregivers the decrease in injuries after the lifts are being used routinely.
- Do not permit manual lifting except in life-threatening circumstances.
- Include caregivers and residents in the selection of lifting equipment.

- Allow caregivers the opportunity to work with different mechanical lifts. Some vendors will allow equipment to be evaluated on a short-term trial basis.
- Ask maintenance and housekeeping staff to provide their opinion and input on the equipment being considered for purchase.
- Ensure that all shifts are covered by an adequate number of caregivers who have been trained to use the lifts to help decrease these injuries.
- Follow-up to check if lifting equipment is being used properly.
- Keep equipment readily available and accessible. The number of lifts required will depend on the level of physical dependency among the residents. As a general rule, one full-body lift should be provided for every eight to ten non-weight bearing residents and one stand-up lift should be provided for every eight to ten partially weight bearing residents (See illustrations of mechanical devices used to transfer residents).
- Provide back-up battery packs on remote chargers as needed so that lifts can be used 24 hours per day while batteries are being recharged.
- Ensure that sufficient slings of the proper size are available.
- Consider a single-patient-use disposable sling for each resident; reimbursement may also be available.
- Store equipment in a convenient location.
- Implement a routine maintenance program to ensure equipment is kept in good working order (the maintenance program should include tagout and repair procedures for broken equipment).
- Provide training to a knowledgeable person with enthusiasm and leadership capabilities on each shift to serve as a peer safety leader. A peer safety leader can provide education, bedside assessments, and training/re-training on lifting equipment.

Mechanical Devices Used to Transfer Residents

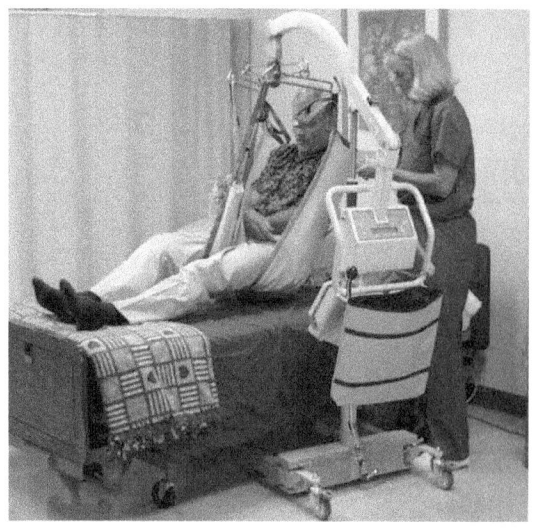 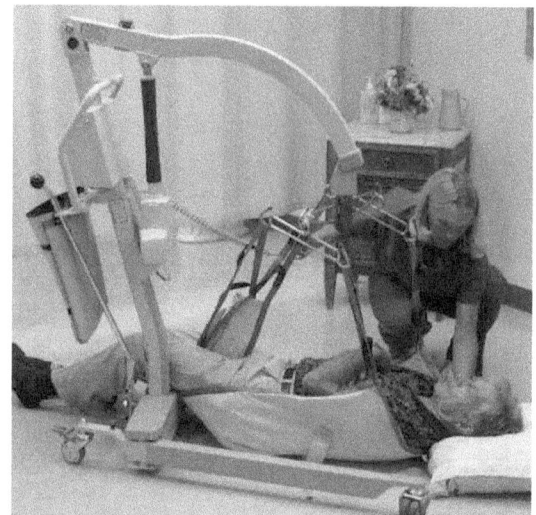

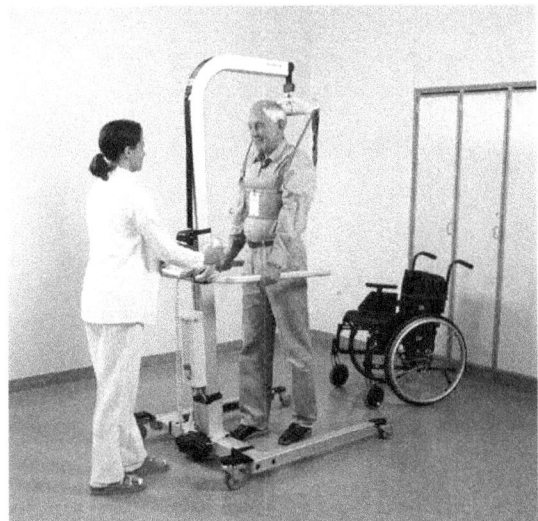

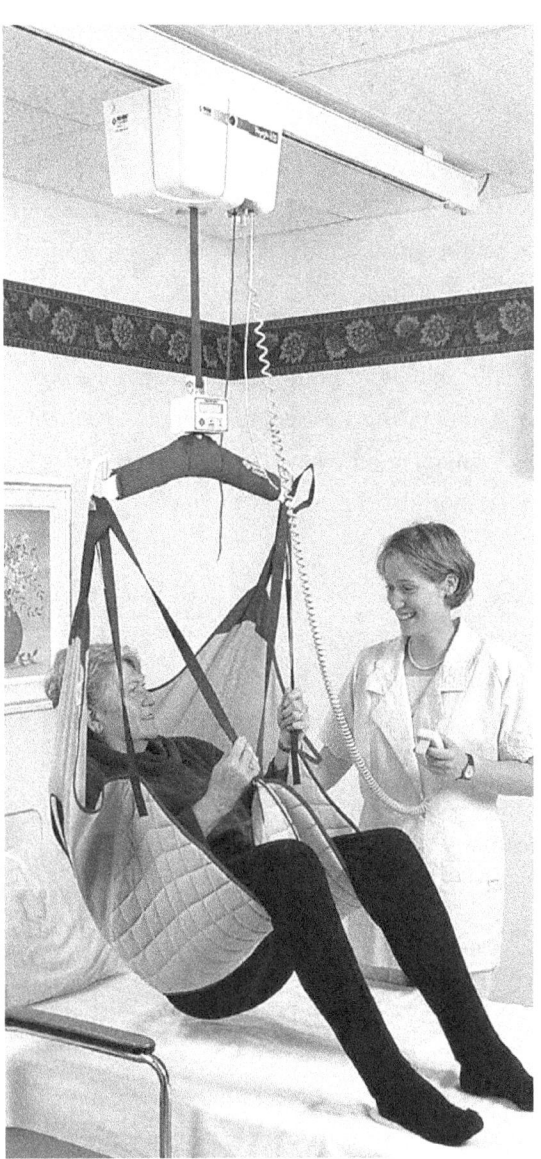

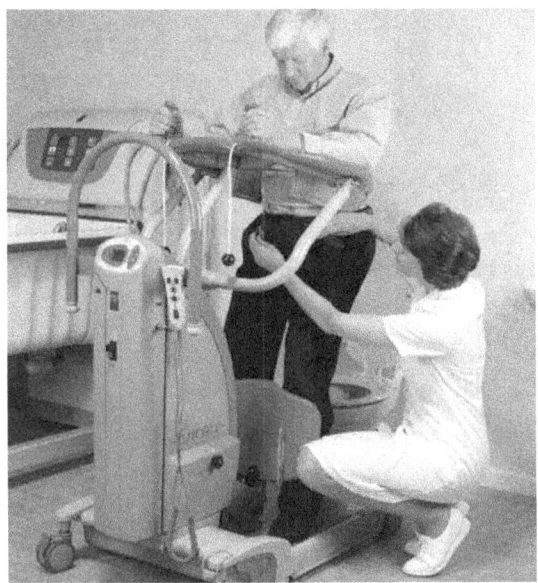

(4) What type and how much training is necessary to ensure that all of the caregivers are prepared to use lifting equipment?

Training should focus on how to use the lifting equipment for residents with a range of physical limitations, and should include hands-on practice. Caregivers should be required to demonstrate that they are proficient in the use of the lifting equipment for residents with a range of disabilities. Training is generally provided by the lifting equipment manufacturer when equipment is purchased; however, a member of the care giving staff and/or peer safety leaders should be trained in all aspects of lifting equipment usage and should be prepared to provide periodic refresher training to existing staff and newly hired caregivers.

(5) Is it helpful to have a written resident lifting policy?

Yes, a written policy establishes:

• Manual lifting is unsafe for residents and staff and is not permitted,
• Minimum standards for the lifting program,
• Transferring needs of each resident are assessed and reassessed as a resident's transferring needs change,
• The amount of lifting equipment required,
• Requirements to select appropriate lifting methods,
• Training requirements for caregivers,
• Responsibilities for all caregivers.

(6) What if a resident refuses to be lifted by a mechanical lift?

Upon admission, it should be explained to the incoming resident that the nursing home has a policy requiring the use of a mechanical lift for non-weight bearing residents. It should be explained that the lift is for the safety of the resident and the caregiver. If caregivers are injured, it will compromise the nursing home's ability to provide quality care. If a resident refuses to be lifted with a mechanical lift, the caregiver, therapy staff, and the social worker should spend extra time with the resident to secure their trust and to help them understand that the lifts increase resident and staff safety.

The social worker, administrator, nurse manager, or therapy staff can intervene with the resident's family by explaining the benefits of lifts for the resident and the caregivers. Offer to demonstrate the lift using a family member, and explain that the use of the lift will not compromise the resident's dignity. Furthermore, the resident's comfort and security may be improved, while reducing his/her risk of injury.

(7) What approaches promote the effective implementation of a safe resident lifting program?

It is important to include caregivers and staff from all departments in the program development. Keeping the staff trained and competent in the use of the mechanical lifting equipment is a key component of a successful program. Lack of compliance may result if newly hired employees do not know how to use the equipment.

Preventive maintenance and prompt repair of improperly functioning equipment will facilitate the use of lifting equipment. Ensure that enough lifts are provided and that they are conveniently stored so that they will be available when needed. Also, slings can be lost, especially if the slings are sent offsite to be laundered. Consider laundering slings on site or use disposable slings. Successful nursing homes establish a mechanism, such as peer safety leaders, to champion the program and provide on-going training, monitor problems, and assist in resolving problems on an ongoing basis. This helps to facilitate staff buy in and consistent usage of the equipment.

(8) What impact will mechanical lifting equipment have on resident assaults to caregivers?

Findings from a NIOSH study (Collins et al., 2004) indicated that resident assaults on caregivers decreased after the implementation of mechanical lifting equipment to lift and transfer residents. After caregivers are sufficiently trained, residents trust that caregivers can move them comfortably and safely. This may partially explain the reduction in assaults from residents. It should be noted that there are exceptions and some residents who are confused or disoriented may violently oppose being lifted with a mechanical lift (Collins et al., 2004).

(9) Does it take more time to use a mechanical lift to move a resident than to manually transfer the resident?

It is quicker to manually transfer a resident. However, using a mechanical lift is much safer for the caregiver and provides a more comfortable and secure transfer for the resident. The long-term health and wellness of the caregiver will be much greater over the long-term by taking a few extra minutes to lighten the daily burden of work. Much of the extra time to use a mechanical lift is spent in locating and bringing the lift to the bedside. Convenient storage and adequate numbers of mechanical lifts greatly reduce the time required to move a resident and increases staff adherence to the program. Ceiling-mounted lifts address the concern of bringing the lift to the bedside because they are conveniently stored in the resident's room.

(10) How should residents be evaluated to assign an appropriate lifting device to the resident?

The first step is to assess the transfer needs of the resident which should be reassessed if the resident's condition changes. The assessment should consider the dependency level of the resident, the size of the resident, the resident's cognitive status, and weight bearing ability. Once the resident's transferring needs have been assessed, it is critical that the information be communicated to the caregivers responsible for lifting the resident. OSHA and VHA documents contain transfer-specific algorithms to help assess the appropriate transferring needs for the resident. These algorithms can be accessed at the Web sites listed in the "More Information Section".

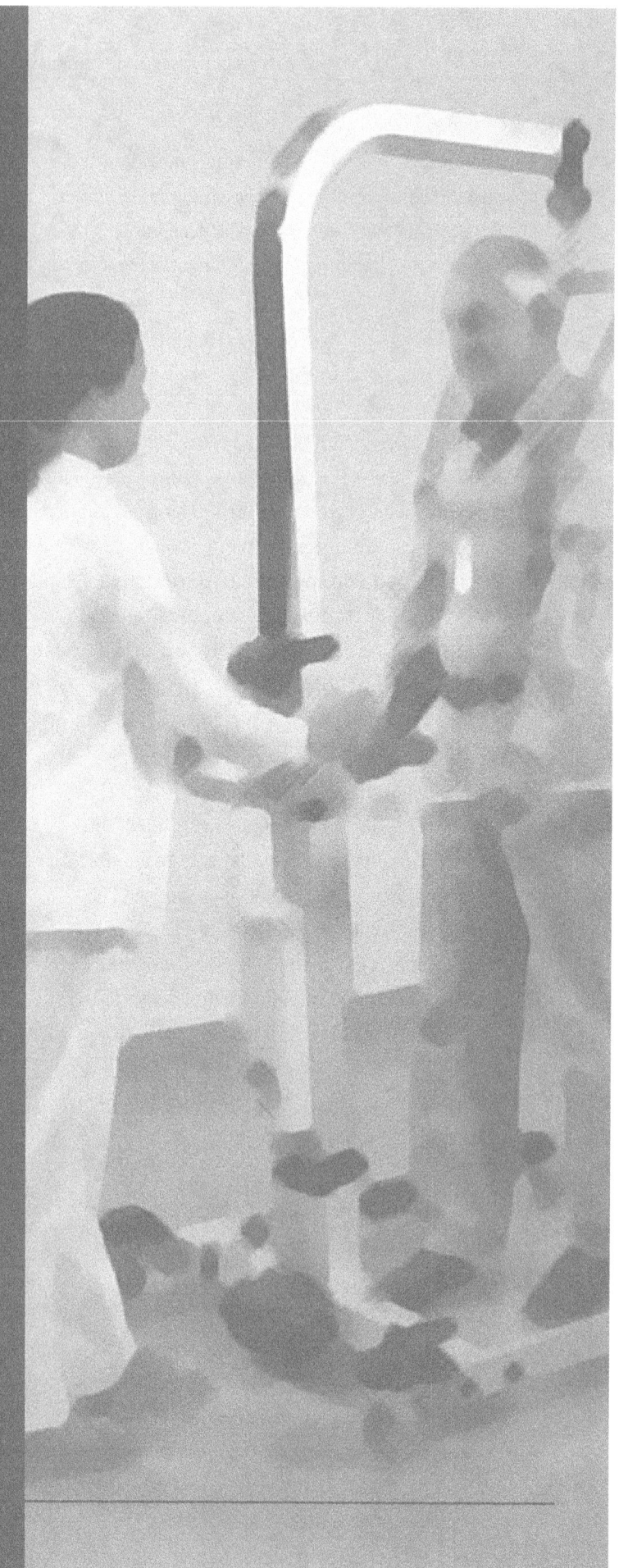

Conclusion

This document provides information for nursing home owners, administrators, nurse managers, safety and health professionals, and workers who are considering establishing a safe resident lifting program. Research has shown that incorporating mechanical lifting devices into a safe resident lifting program decreases caregiver injuries, lost workdays, workers' compensation costs, and employee turn-over while improving employee morale and the quality of care for residents. The initial investment in the equipment and training is quickly recovered because of the reduced injury costs to caregivers.

More Information about Setting Up a Safe Resident Lifting Program Using Mechanical Equipment can be Found at:

Patient Care Ergonomics Resource Guide: Safe Patient Handling and Movement, Veterans Administration Hospital, Tampa, Florida and Department of Defense, 2001. www.patientsafetycenter.com/Safe%20Pt%20Handling%20Div.htm

OSHA, Guidelines for Nursing Homes: Ergonomics for the Prevention of Musculoskeletal Disorders, OSHA 3182, 2003. www.osha.gov/ergonomics/guidelines/nursinghome/

Handle with Care: The American Nurses Association's Campaign to Address Work-Related Musculoskeletal Disorders. http://www.nursingworld.org/handlewithcare/brochure.htm

References

Collins JW, Wolf L, Bell J, Evanoff B [2004]. An Evaluation of a "Best Practices" Musculoskeletal Injury Prevention Program in Nursing Homes. Injury Prevention 10:206-211.

Department of Health and Human Services, Public Health Service, Centers for Disease Control and Prevention, National Institute for Occupational Safety and Health [1994]. Workplace Use of Back Belts – Review and Recommendations, NIOSH, Pub No. 94-122, 1994.

Garg A [1999]. Long-Term Effectiveness of "Zero-Lift Program" in Seven Nursing Homes and One Hospital, Contract No. U60/CCU512089-02. http://www2.cdc.gov/nioshtic-2/Nioshtic2.htm.

Garg A, Owen BD, [1992]. Reducing Back Stress to Nursing Personnel: An Ergonomics Intervention in a Nursing Home. Ergonomics 35:1353-1375.

Lagerstrom M, Hansson T, Hagberg M [1998]. Work-Related Low Back Problems in Nursing. Scandanavian Journal of Work and Environment and Health 24(6):449-464.

Nelson A, Fragala G, Menzel N [2003a]. Myths and Facts about Back Injuries in Nursing, American Journal of Nursing, 103 (2):32-40.

Nelson A, Matz M, Chen F, Siddharthan K, Lloyd J, and Fragala G [2003b]. Research Report: A Multifaceted Ergonomics Program to Prevent Injuries Associated with Patient Handling Tasks in the VHA.

Occupational Safety and Health Administration Web site: http://www.osha.gov/ergonomics/guidelines/nursinghome/intro2.html

Tiesman H, Nelson A, Charney W, Siddharthan K, Fragala G [2003]. Effectiveness of a Ceiling-mounted Patient Lift System in Reducing Occupational Injuries in Long Term Care. Journal of Healthcare Safety 1 (1): 34-40.

U.S. Department of Labor Bureau of Labor Statistics [2003].Total Recordable Occupational Injury Cases in Nursing and Residential Care Facilities. Available at http://www.bls.gov/data/home.htm (Accessibility verified 02/04/05)

U.S. Department of Labor, Bureau of Labor Statistics [2004]. February 2004 Monthly Labor Review, Table 4 – Occupations with the largest job growth, 2002-2012. Available at http://www.bls.gov/emp/emptab4.htm (Accessibility verified 07/21/05).

Wassell JT, Gardner LI, Landsittel DP, Johnston JJ, Johnston JM [2000]. A Prospective Study of Back Belts for Prevention of Back Pain and Injury. Journal of the American Medical Association 284(21):2727-2732.

Zhuang Z, Stobbe TJ, Collins JW, Hsiao H, Hobbs, G [2000]. Psychophysical Assessment of Assistive Devices for Transferring Patients/Residents. Applied Ergonomics 31:35-44.

Safe Lifting and Movement of Nursing Home Residents (NIOSH Pub. No. 2006-117)

Please take a few minutes to help us evaluate the enclosed NIOSH guide. Fold, tape ends together, and mail when completed. Thank you.

1. **Are you involved in the operation of a nursing or personal care home, or in the lifting or movement of nursing home residents?** ☐ yes ☐ no

1.a. If yes, check the item(s) below that best describes your involvement:

☐   A nursing or personal care home owner

☐   A nursing or personal care home administrator

☐   A manager of nursing services

☐   An occupational safety and health specialist

☐   A physician

☐   A physical therapist

☐   A nurse, aide, or orderly

☐   Other, please specify: _____

Proceed to question 2.

1.b. If no, please describe your role or interest in the issue of safe resident lifting in nursing and personal care homes:

_____

Proceed to question 3.

2. **Which of the following best describes the resident lifting program in your facility?**

☐   The facility has no resident lifting program

☐   The facility has a safe resident lifting program that includes (check all that apply):

    ☐   a written resident lifting policy, procedures

    ☐   the use of mechanical lifting devices

    ☐   staff training in use of mechanical lifting devices

    ☐   other components (specify: _____)

3. **How did you read this guide?**

☐   Cover to cover

☐   Selected sections only (please specify all that apply):

    ☐   Introduction

    ☐   The Challenge of Lifting Residents in Nursing Homes

    ☐   Benefits, Cost, and Effectiveness of a Safe Resident Lifting Program

    ☐   Frequently Asked Questions

    ☐   Conclusion

    ☐   More Information

    ☐   References

    ☐   Did not read

Was the information used (check all that apply)?

☐   to pass along to/inform someone else

☐   for communicating to managers or staff

☐   to obtain further information on this issue

☐   to review/evaluate current facility policies and programs

☐   to change current facility policies and programs: (specify: _____)

_____

_____

_____

☐   to implement the following activities or components of a resident lifting program:

    ☐   develop and communicate a written policy

    ☐   invest in portable (not ceiling mounted) mechanical lifting devices

    ☐   invest in ceiling mounted mechanical lifting devices

    ☐   train staff in the use of mechanical lifting devices

    ☐   to implement a comprehensive resident lifting program based upon the guide

    ☐   the information was <u>not</u> used

Other comments: _____

_____

_____

_____

www.ingramcontent.com/pod-product-compliance
Lightning Source LLC
Chambersburg PA
CBHW081757170526
45167CB00009B/4049